Computational fluid dynamics in building design

An introductory guide

Richard Chitty and Chunli Cao

The research and writing for this publication has been funded by BRE Trust, the largest UK charity dedicated specifically to research and education in the built environment. BRE Trust uses the profits made by its trading companies to fund new research and education programmes that advance knowledge, innovation and communication for public benefit.

BRE Trust is a company limited by guarantee, registered in England and Wales (no. 3282856) and registered as a charity in England (no. 1092193) and in Scotland (no. SC039320).
Registered office: Bucknalls Lane, Garston, Watford, Herts WD25 9XX
Tel: +44 (0) 333 321 8811
Email: secretary@bretrust.co.uk
www.bretrust.org.uk

IHS (NYSE: IHS) is the leading source of information, insight and analytics in critical areas that shape today's business landscape. Businesses and governments in more than 165 countries around the globe rely on the comprehensive content, expert independent analysis and flexible delivery methods of IHS to make high-impact decisions and develop strategies with speed and confidence. IHS is the exclusive publisher of BRE Trust publications.
IHS Global Ltd is a private limited company registered in England and Wales (no. 00788737).
Registered office: Willoughby Road, Bracknell, Berkshire RG12 8FB. www.ihs.com

BRE Trust publications are available from www.brebookshop.com
or
IHS BRE Press
Willoughby Road
Bracknell
Berkshire RG12 8FB
Tel: +44 (0) 1344 328038
Fax: +44 (0) 1344 328005
Email: brepress@ihs.com

Printed using FSC or PEFC material from sustainable forests.

FB 69

First published 2014

ISBN 978-1-84806-374-7

Contents

Cont'd . . .

Case studies

Abbreviations

ABL Atmospheric boundary layer

ACH Air changes per hour

AIAA American Institute of Aeronautics and Astronautics

ASTM American Society for Testing and Materials

CAD Computer-aided design

CFD Computational fluid dynamics

DES Detached eddy simulation

DNS Direct numerical simulation

DOM Discrete ordinates model

DTRM Discrete transfer radiation model

FDS Fire dynamics simulator

FEM Finite element modelling

FSI Fluid–structure interaction

HVAC Heating, ventilation and air conditioning

ISO International Standards Organization

LES Large eddy simulation

NIST (United States) National Insitute of Standards and Technology

OUT_SET Outdoor standard effective temperature

PET Physiological equivalent temperature

PMV Predicted mean vote

PPD Percent persons dissatisfied

RANS Reynolds-Averaged Navier-Stokes

1 Introduction

Computational fluid dynamics (CFD) modelling is a powerful tool that is routinely used in fire engineering and building design[1,2]. It can be used to simulate different aspects of a building's performance or its impact on the local environment.

This guide gives some of the background to the CFD methodology for non-experts such as Building Control and other regulatory authorities who have to approve or accept a design. They may be presented with the visualised results of CFD simulations in various graphical formats (including animations) that show the predicted performance. However, underlying assumptions, factors and limitations that impact on the quality of the predictions tend to be buried in detailed technical reports.

The process of reviewing CFD predictions is often dependent on the ability of the reviewer to be able to identify 'schoolboy' errors (eg supersonic wind velocities or fire temperatures of 5000 °C!) using simple sanity checks based on their experience of the real world rather than the mathematical limits of the simulation environment. Visualisation methods often mask the difference between:

- a *qualitative* simulation to demonstrate the feasibility of a concept
- a *quantitative* simulation required to prove functionality of different aspects of the design.

This guide discusses the current capabilities and limitations of CFD for building design and gives guidance on best practice, focusing on:

- fire safety
- ventilation
- thermal comfort
- wind movement around buildings.

2 Computational fluid dynamics explained

The movement of air and other gases (such as smoke from fires, fuel gases or releases of pollutants) around, into and through buildings can be complex. These flows can be driven by wind, buoyancy or by mechanical means such as heating, ventilation and air conditioning (HVAC) systems incorporating fans, and often a combination of all three. In building design or the assessment of building performance, these flows may need to be predicted. In many cases, prescriptive rules such as those given in the technical guidance associated with the national building regulations may be sufficient (eg ventilation rates for car parks to remove exhaust fumes). However, if more detail is required (eg velocity and temperature distributions in an office space) then some form of physical or numerical simulation is required.

This guide covers the application of numerical (computer) simulations and especially the use of computational fluid dynamics (CFD) and the presentation of the simulations. Physical simulations using wind tunnels or environmental chambers are discussed elsewhere (for example, BRE Digest 520[3]).

Mathematical descriptions of the conservation of mass, momentum and energy in flow of fluids (either liquid or gaseous) have been used to develop sets of equations whose solutions would describe the velocity, pressure, temperature, etc. at any location in the fluid for a given set of boundary conditions. Unfortunately, these equations are difficult to solve and require simplifying assumptions and, usually, numerical solution methods. Computing these solutions is an ongoing challenge to the largest and fastest of computers.

The underlying principle of CFD is the conservation of properties in the flow. This can be considered as:

Change of quantity in a space = flow in/out of the space
+ diffusion in/out of the space
+ generation or loss in the space.

Or mathematically as:

$$\frac{\partial \rho\phi}{\partial t} = \frac{\partial U_i \rho\phi}{\partial x_i} + \frac{\partial}{\partial x_j}\left(\Gamma_\phi \frac{\partial \phi}{\partial x_i}\right) + S_\phi \qquad \text{(Eqn 1)}$$

where:

ϕ = the quantity being considered (e.g. mass, momentum, concentration, energy)

ρ = the density of the fluid

U_i = the velocity in the x_i direction

t = time

Γ_ϕ = the diffusion coefficient for the quantity being considered (eg conduction for energy)

S_ϕ = the 'source term' describing the amount of the quantity being created or lost.

Being able to use the same basic equation for each quantity being considered is attractive for a computer solution as one efficient solution technique can be developed and used repeatedly for each quantity.

The solution approach requires approximating the differential terms into an algebraic form. For a simple case:

$$\frac{dy}{dx} = \frac{Y_2 - Y_1}{\Delta X} \qquad \text{(Eqn 2)}$$

Where Y_1 and Y_2 are the values a distance ΔX apart. This discretisation process, which is much more complex for the CFD equations, is an approximation which has increasing accuracy with smaller distances. However, a small step size implies the space has to be divided into a large number of intervals which increases the computational time and memory requirement. In practice, more advanced discretisation schemes than that given in Equation 2 are used to improve accuracy, numerical stability and efficiency of the solution procedure. A similar discretisation process is applied to time.

Solving Equation 1 at a number of points in space and time introduces the concept of the CFD mesh or grid. A very fine grid may produce accurate results but it will have a large number of cells which will require a long computation time. One of the significant challenges in using the CFD technique is the creation of a grid that provides both an accurate and reliable solution in an acceptable computation time.

Equation 1 does not include all the phenomena that may need to be simulated. For example, a chemical reaction, such as combustion, can be included by using the source term to remove reactants and create products. However, further calculations would be required to determine the actual values used in the CFD equation. This requires a sub model to represent the chemical reaction.

Other important sub-models include turbulence, radiation and heat transfer to a surface. Commercial CFD software will include libraries of sub-models for different applications and provision for users to create their own. Open source software such as OpenFOAM[4] has a large number of sub models created by a user community for a wide range of applications. Specialist CFD software, such as the fire dynamics simulator (FDS)[5], include sub-models relevant to their use which are selected using the terminology relating to the application, thereby providing a user-friendly interface.

The final requirement to solve Equation 1 is a set of boundary conditions. These are fixed solutions of Equation 1 at various locations and times (usually at the edge of the grid and when time = 0). In physical terms, these represent part of the problem description such as the flow rates from fans (a source term S_ϕ) or an initial temperature. It will not be possible to obtain a solution to a CFD problem if there are too few boundary conditions (the problem is under prescribed) or if there are too many (the problem is over prescribed). A simple example of an over-prescribed problem is a pipe containing a fluid with a constant density flow, if the velocities at each end of the pipe are set to different values, the inflow and the outflow will never match, Equation (1) can never be solved and the numerical methods used by computer software will not converge to a single solution. This is a key issue that any user of a CFD code should be aware of.

2.1 Grids

The simplest grids are Cartesian meshes which are rectangular blocks divided into a number of smaller rectangular cells as shown in Figure 1. These cells do not have to be uniformly distributed and can be spaced so that more cells are present where the flow is expected to vary rapidly.

The building geometry (eg walls or roofs) and other objects such as furniture can be represented by blocking some of the cells to prevent the fluid flow as illustrated in Figure 2. This is often adequate for simulations in buildings where rooms are often rectangular. However, sloping surfaces such as a pitched roof would have to be approximated by a series of steps. To efficiently represent more complex geometries, several grid blocks may be combined (multi block meshes) as shown in Figure 3. It is also possible to distort a rectangular block to include some curved or sloping features as can be seen in Figure 4.

For more complex shapes, unstructured meshes are used. These use cells of various shapes (tetrahedral, hexahedral and pyramids) and can accurately and efficiently represent complex shapes.

Different programs will have different grid options and may impose restrictions on the grid specification (for example, FDS requires the grid to be linear in at least one direction and requires the number of cells to be in the form $2^i3^k5^j$, this is a constraint imposed by the high performance mathematical routine used to solve Equation 1). The CFD software may include specific tools to assist the user in creating the CFD grid.

As the grid used is an artificial artefact for the numerical method, the solution should be independent of the grid used. If the grid is coarse, averaging in the cells will remove fine details of the flow which may have the effect of altering the computational solution. A grid sensitivity analysis should be performed to demonstrate a particular solution is not dependent on the grid that has been used.

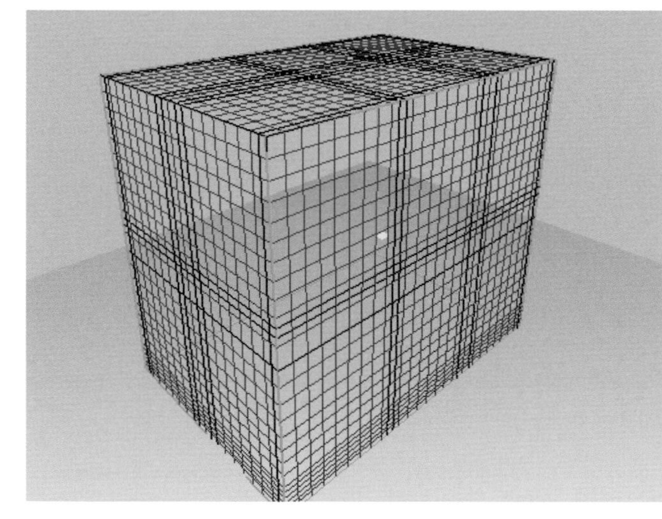

Figure 1: Simple structured mesh

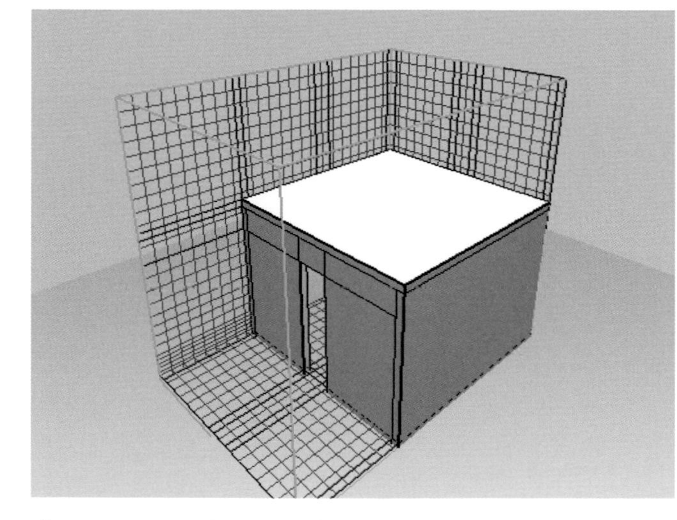

Figure 2: Structured mesh with blocked cells

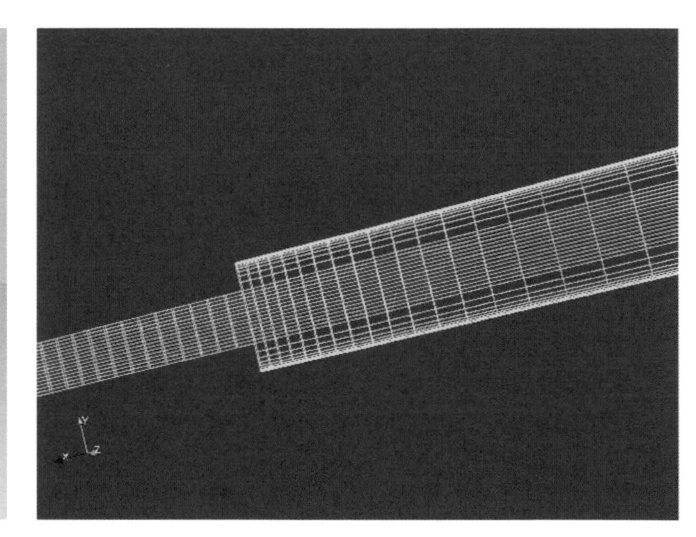

Figure 3: Multi-block grid

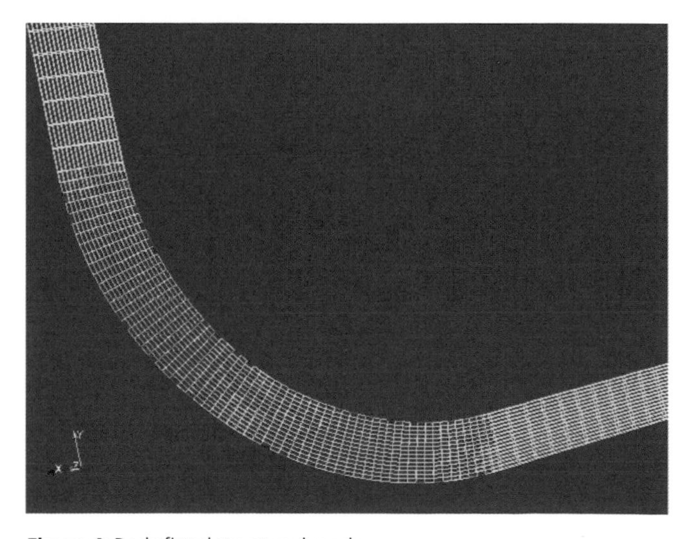

Figure 4: Body fitted structured mesh

2.2 Sub-models

Detailed discussion of the basic CFD methodology and sub-models can be found in a number of text books, for example, *An introduction to computational fluid dynamics: the finite volume method*[6].

2.2.1 Turbulence

It is not usually practical to make a grid that is fine enough to capture every movement in a flow. However, it is important that the kinetic energy in the smaller eddies is included in the calculations and this is achieved using a sub model. A wide range of sub-models has been developed, some are general-purpose models intended to be used with a wide range of flows, others are more specialised for particular problems (eg supersonic transitions). For building applications, some of the more general models can be used. However, they need to be selected with care to ensure that they are appropriate for the specific problem such as wind flows where a boundary layer can lift off the top of a building.

Relevant turbulence models are discussed in more detail in later sections of this guide within the context of the specific applications.

2.2.2 Thermal radiation

CFD models predict the transport of heat by convection, ie heat being *carried* by the moving gas. Heat transfer by thermal radiation is a different mechanism where the energy is transported by electromagnetic radiation along a *line of sight* at the speed of light. At low temperatures it may be possible to justify neglecting thermal radiation but at the temperatures encountered in fires, thermal radiation is significant and may be the dominant heat transfer mechanism.

Thermal radiation is included in CFD simulations using sub-models. Some of the more common radiation models in use include:

- discrete ordinates model (DOM)
- discrete transfer radiation model (DTRM)
- flux model.

Of these, the DOM is the most comprehensive. It includes absorption, scattering and emission but at the cost of additional computation time. The DTRM uses ray tracing to integrate radiant intensity along a ray that is considered to be representative of a solid angle at a surface. It does not include scattering, but applies to a wide range of optical thicknesses.

The accuracy of the DTRM increases with the number of rays, but as with the DOM this also increases the computation time. Flux models are much simpler but only consider radiation in a limited number of directions. As such, they tend to be less accurate but can be formulated into the form of Equation 1 so they can be solved using the CFD numerical solver.

In some building applications, the optical properties of the fluid (air) may be constant throughout the space and simply represented. However, in fire simulations where there may be products of combustion (smoke and chemical species such as carbon dioxide and water vapour) in the radiation path, a more detailed approach is required.

2.2.3 Combustion

For simulations of a problem involving fires the combustion process has to be considered. The simplest approach is to simply consider a fire as a source of energy (which leads to elevated temperatures) and an effluent to represent smoke and fire gases (eg carbon dioxide). This does, however, have limitations as it requires the flaming region to be prescribed and the fire would not be influenced by factors such as the local oxygen concentration.

A more rigorous approach is to include a combustion model where the reaction rate (and hence the heat release rate and production rate of combustion products) is related to the local fuel and oxygen concentrations and flow conditions. This may be a basic one-step reaction where 'fuel' is converted to 'product' if there is sufficient oxygen or may be more complex and include several reaction steps. Including further reaction steps allows the model to predict the production of partially oxidised products (such as carbon monoxide or oxides of nitrogen), though this is at the cost of further computational effort. There may also be difficulty in acquiring reliable input data for the more advanced models (for example, activation energy parameters for the different reaction steps).

2.3 Other models

In addition to sub-models that are directly required to provide the solution of the flow field, other models may be required to modify the boundary conditions. These include heat transfer from the gases to solid surfaces, pyrolysis models to predict the production rate of combustible gases from a heated liquid or solid and variation of the flow rate of a fan due to a pressure difference between two locations (to simulate a fan and duct work).

3 Computational fluid dynamics in use

3.1 Resources

3.1.1 Software and computers

Until relatively recently (ie the last 10 years) CFD simulation could not be performed without access to what would, at the time, be regarded as high-performance computers. Now (in 2014) readily available desk top personal computers and laptops are capable of performing simulations. The time required to complete the simulation will be reduced with numbers of processing cores, additional memory or networks of servers. However, simulations can take a long time to run and, depending on the complexity of the problem, may extend to several days or weeks. To ensure that simulations can provide solutions efficiently, it is best practice to save 'restart files' frequently and use uninterruptible power supplies.

There are a number of computer software products available to perform CFD simulations. These may be free, open source codes (eg OpenFOAM[4] and FDS[5]) that originate from the research community and are supported, to various degrees, by the developer and/or a user community. There are also commercial codes (eg PHOENICS[7] and ANSYS CFX[8]) which are supplied with support contracts and training courses at various levels.

3.1.2 Verification and validation

Using computer simulations, especially for life safety applications such as fire safety in buildings, raises questions regarding the reliability and accuracy of the simulations. This can be addressed, in part, by considering verification and validation of the software. Simplistically, verification confirms that the equations being solved by the computer are providing the correct mathematical results (within an accepted level of accuracy) and validation confirms that, for a given scenario the simulation provides a credible representation of reality. Organisations such as ISO, AIAA and ASTM have developed standards and guidance for the demonstration of verification and validation of CFD and other simulation models relevant to building design (eg BS ISO TR 13387-3[9]).

In the context of fire safety, the US Nuclear Regulatory Commission has conducted an extensive verification and validation of some widely used fire simulation models[10], including FDS[11].

3.1.3 Time

A major resource required for a CFD simulation is time. However, it is important to realise that much of this may be computation time which may not have a big impact on project costs. A CFD simulation is not necessarily an expensive option. A significant amount of effort may be required to create the model geometry by building a grid, determining material properties, selecting an appropriate combination of sub models and identifying values for solution control parameters. This would usually involve several trial runs and include an investigation of grid sensitivity. Some of the software, especially the commercial codes, will provide tools (eg interfaces allowing the use of computer-aided design [CAD] files) to assist in this process.

3.1.4 Grid refinement

As a rule of principle, finer meshes give more accurate predictions. However, in reality, there is usually pressure to use coarser meshes to save computer resources and provide an answer more quickly. There should always be a cautious approach to ensure that a mesh has a high enough density to give predictions to the required accuracy. A good approach is to perform trial runs using successively finer meshes until quantitative values in key areas (such as the flow rate of air or smoke through a door, or a temperature at a particular location) do not change significantly when the mesh density is increased.

3.2 Analysis of results

CFD simulations can create large quantities of data (100s of megabytes) and analysis can be a significant task. In some cases, the 'required' output maybe a single number, for example, 'what time would a heat detector at a specific location detect a fire?'. However, CFD simulations can reveal flow conditions that have not been anticipated (which might make it possible to use simpler models that are less expensive to run) and therefore the results need to be carefully examined and compared (validated) with results from simpler calculations and experimentally characterised conditions or using 'engineering judgement'. To continue the heat detector example, if the simulation shows a rapid response then this would usually be regarded as a 'good' outcome. However, the reason for the rapid response should be examined as it may be an artefact of a fortuitous arrangement of the fire, ventilation conditions and detector location. A small change in the location or characteristics of the fire (usually one of the uncertainties in the input data for a fire simulation) could result in the fire not being detected at all!

Examining the predicted flow fields using the graphical tools available for CFD programs can confirm anticipated flows and reveal any unusual features.

3.2.1 Graphical analysis

Most CFD software includes programs for displaying the geometry, CFD grid and results of simulations in a number of ways. Commercial products tend to have fully integrated systems where the user creates a project and can then easily manipulate and exchange data, possibly conducting several types of simulation (eg CFD and structural simulations) as the work progresses. There is independent data analysis software, such as FIELDVIEW and ParaView, that can import results using customised 'plug-ins' or standard file formats to read data from different CFD programs.

The most familiar graphical methods for showing CFD data are vector and contour plots as shown in Figures 5 and 6. These are frequently used on weather maps to show the speed and direction of wind (vector plots) and temperature or rainfall distributions (contour maps).

Another graphical method, streamlines, can be used to show the flow over surfaces and around or through objects, but these can be confusing with re-circulating flows as illustrated in Figure 7.

These displays of streamlines and vectors can be elaborated, as in the figures, by setting the colour depending on the value of another variable such as temperature. Streamlines can also be drawn as ribbons that twist indicating any swirl component in the flow.

A further option is to create iso surfaces which are surfaces that enclose regions where a variable is greater than or equal to a selected value (Figure 8).

In fire simulations iso surfaces are often used to show the location of a hot gas layer containing smoke. However, it should be noted that this does not necessarily correspond to the actual visual impact of the scenario, but it can clearly show the extent of a hazard in a building. The iso surfaces shown in Figure 8 have

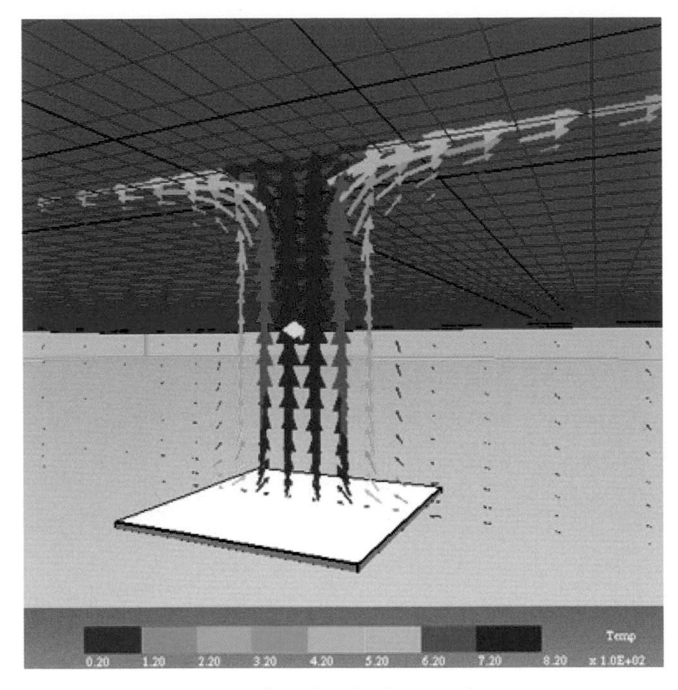

Figure 5: Vectors showing flow direction (JASMINE)

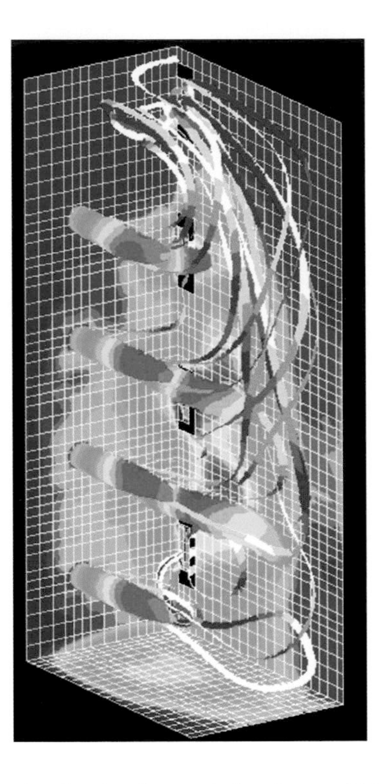

Figure 7: Streamline showing flow paths in a furnace (SOFIE/FieldView)

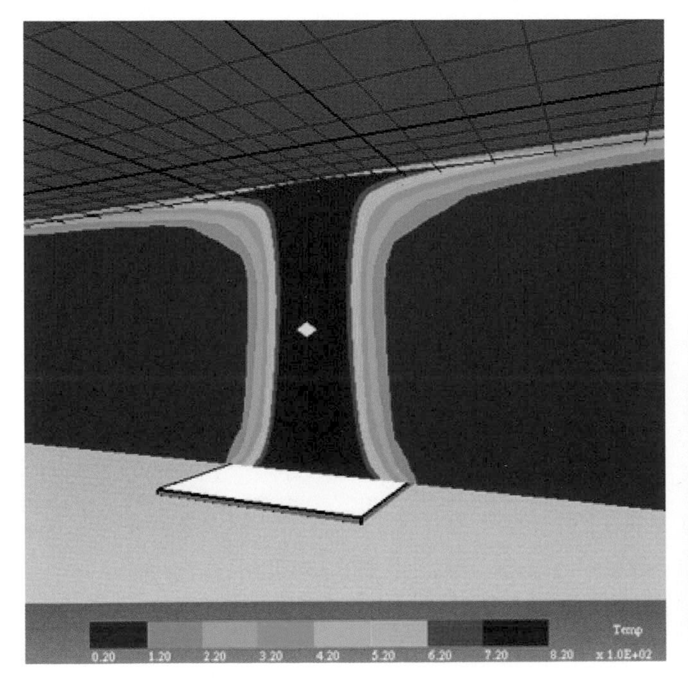

Figure 6: Contour map showing temperatures (JASMINE)

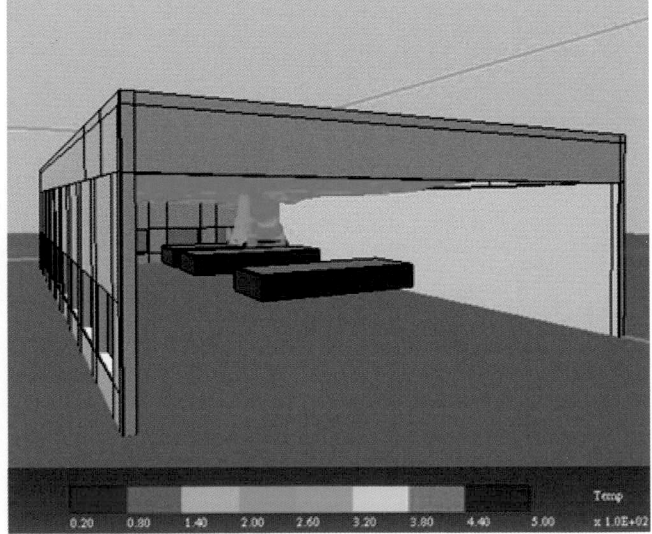

Figure 8: Iso surface

been shaded in green to indicate that it does not represent the actual appearance of the smoke in the enclosure.

A combination of methods can be used to provide a good general impression of the simulation results. Figure 9 shows effluent being discharged from stacks on a building.

Visualisation of CFD results can appear complex. It is important that complexity is not confused with accuracy or reliability of solutions and output. However, understanding of the outputs is much easier if the context of the visual image is clear through the inclusion of identifiable features (eg furniture in a building) to convey a sense of scale and orientation. However, adding realistic features to images including simulation results should not be done without explanation as it may be assumed that the features were included in the simulation as shown. For example, the chairs in Figure 10 were not included as obstructions in the simulation but are just included in the visualisation as an indication of scale and context.

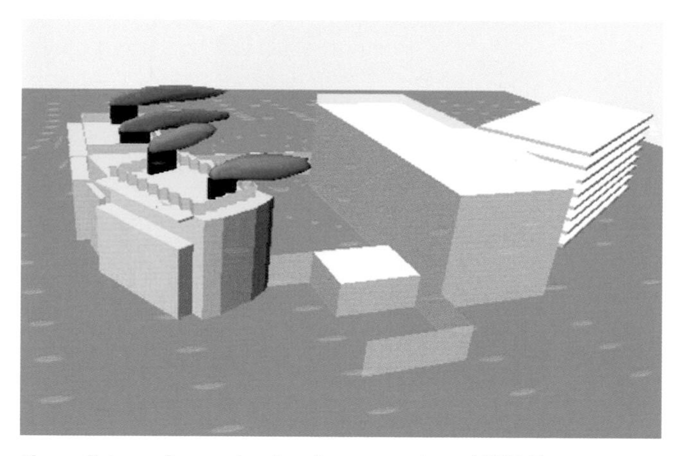

Figure 9: Iso surface and vectors (image courtesy of BISRIA)

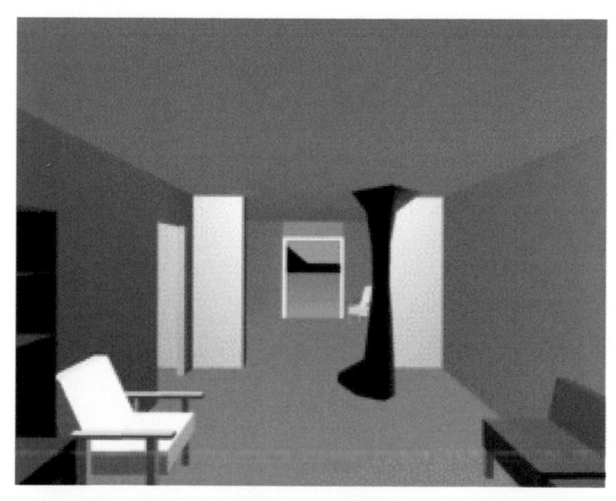

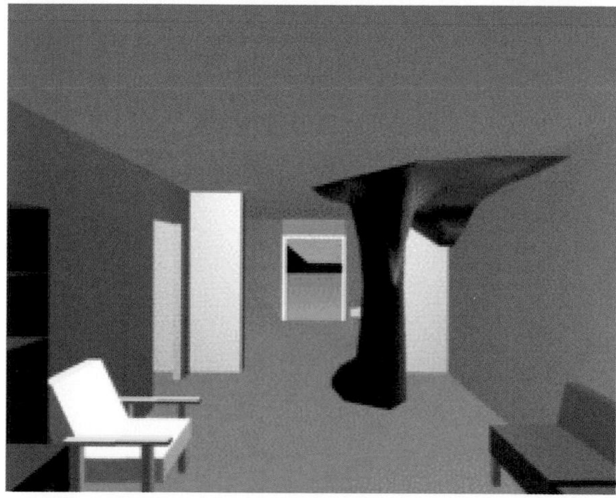

Figure 10: Iso surface showing smoke layer development in a room

3.2.2 Animation

Animated displays either allow the time component of a solution to be examined or allow the observer to move around the simulation to view the geometry and data from different positions.

The results of some simulations, such as those using large eddy simulation (LES) turbulent models may be misleading if only viewed as a still image as the data at some locations can change rapidly with time (see, for example, Figure 11).

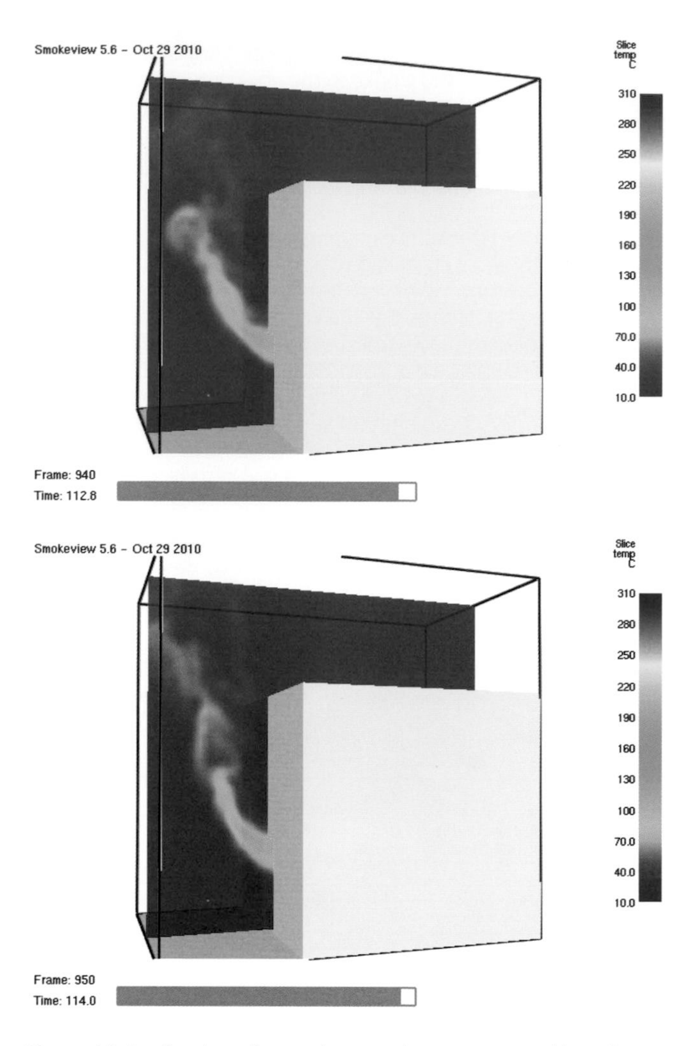

Figure 11: Smoke plume from a doorway, images separated by a short time

Building applications

4 Fire safety design

CFD has become an established tool in the fire safety design of buildings to demonstrate that various systems will be adequate, in the event of a fire to enable the evacuation of the occupants of a building and provide safe access for firefighters. Their use is outlined in general in BS 7974:2001 *Application of fire safety engineering principles to the design of buildings. Code of Practice*[12] and more specifically in Standards such as BS 7346-7:2006 *Code of practice on functional recommendations and calculation methods for smoke and heat control in covered car parks*[13].

CFD is not a design tool that will calculate fan capacities, vent sizes, etc. What it will do is simulate the conditions in a space having a specified fire and fire protection and control measures with given performances. The conditions can be assessed using criteria that are deemed to be relevant for the particular situation. CFD simulations can be used to predict temperatures, gas velocities and species as a function of time. The results can be used to predict:

- smoke movement
- detector operation times
- hazards to occupants
- hazards to materials (fire spread)
- hazard to the structure
- radiation intensity
- possibility of fire spread.

These predictions can include the operation of fans and the presence of smoke barriers. Some CFD programs may also include heat transfer to water sprays (eg sprinklers) and flame spread.

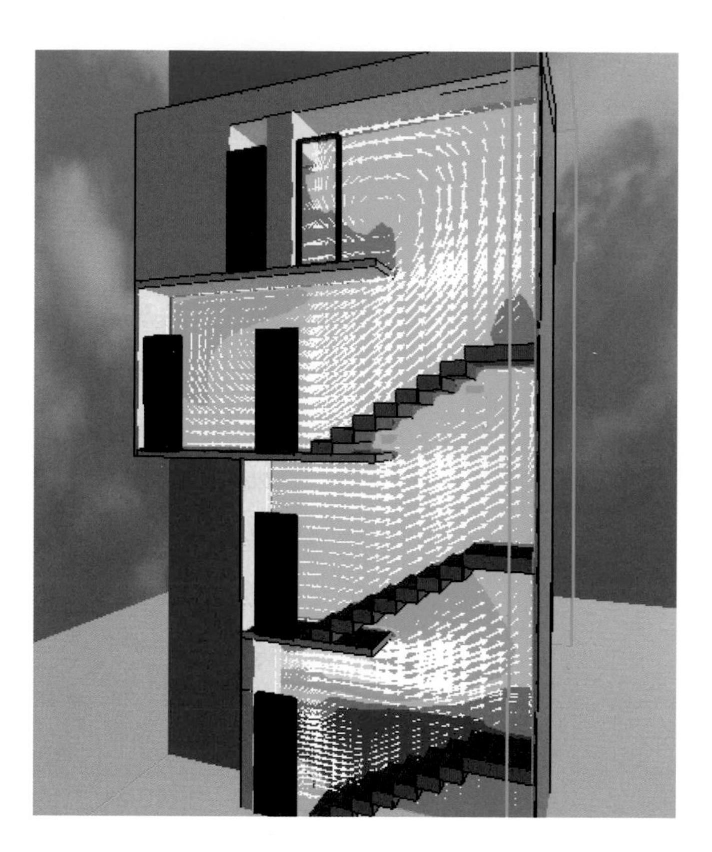

4.1 Data input

The key input parameter to any fire simulation is the heat output and location of the fire. In a design exercise, BS 7974:2001[12] suggests a qualitative design review where relevant hazards are identified by members of the design team led by a fire safety engineer. This creates a specific objective and relevant criteria for the design, including the definition of a credible fire scenario. This can then be quantified using simulation tools such as CFD and the results compared with appropriate criteria to establish the success or failure of the system.

In addition to the design fire scenario, the detailed geometry of the building is required. This will usually be simplified (small features can usually be removed) and restricted to a small part of the building relevant to the problem being examined.

Material properties are required so that heat transfer into the structure can be included. Some CFD programs allow the material properties to be dependent on temperature.

Finally, criteria need to be determined for the activation of fans and opening of doors, windows and vents.

Some of the input data may need to be determined from small-scale tests (eg the cone calorimeter) or published experimental data.

4.2 Data output

The most obvious outputs from a CFD simulation of a fire problem are predictions of temperature and smoke concentration as functions of time and location. There will be predictions of gas flow rates and direction. In addition, there may also be predictions of major gas species, thermal radiation and temperatures of exposed surfaces. These data will be available at any (or all) of the cell locations and at each calculation time step.

As with all CFD simulations, there is the potential to generate large amounts of data. The software will usually have the capability to store the whole data set at specified times and a subset of the data at each calculation time step. Having a clear objective for the simulation will ensure data are stored economically without the need to re-run a simulation so that additional data items can be extracted at a later date.

It is best practice to anticipate supplementary analysis of the data and save (relatively) small amounts of additional information. For example, in addition to saving data at head height on a particular floor of a building (to establish time to hazard on that floor), storing data at head height on all floors may be useful if the results of a simulation suggest using a phased evacuation strategy (where times to hazard on each floor may be required). If in doubt save the data, as they can easily be archived or deleted but could take several days to recalculate!

4.3 Post processing

The data from a CFD fire simulation can be used to evaluate a number of things that are relevant to a fire safety design, including the data applications discussed below.

4.3.1 The response of thermal detectors

Thermal detectors such as fusible links or sprinkler bulbs do not operate at the point when the local gas temperature reaches the specified operating point but when the temperature of the sensing element reaches that temperature. This effect can be included by introducing an additional calculation relating the temperature of the sensor to the local gas temperature and velocity.

4.3.2 Smoke obscuration along a line of sight (visibility, beam detectors)

Another application of data from a CFD simulation of a fire is the estimation of the visibility in a building. This may either be to assess occupant safety or to determine the operation of an optical beam detector.

4.3.3 Fractional effective dose (toxicity)

While most smoke control systems in buildings would be designed to prevent occupants coming into contact with smoke and hot gases from a fire, there are situations where an estimate of the toxicity of a fire atmosphere is required (for example, where the occupant of a room cannot leave without assistance). The cumulative concentration of heat, oxygen, carbon dioxide and other fire products can be used to estimate a fractional effective dose which will indicate when a person at a particular location will be incapacitated; this will indicate the time available to provide assistance.

4.4 Modelling assumptions

Some of the processes involved in a fire are complex and cannot be rigorously included in a CFD simulation either because the process is not well enough understood to create a sub-model or because the sub-model would be so complex or computationally demanding that it would not be cost effective. In these cases, CFD simulations resort to either simple or empirical models that capture the basic process but may only be valid for a narrow range of conditions.

Some of the sub-models used in fire simulations are discussed in the following sections.

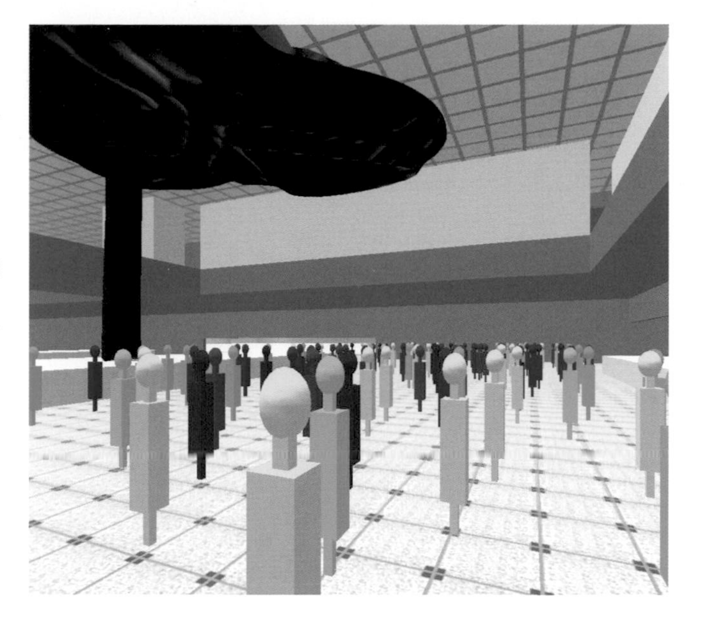

4.4.1 Combustion

Even for simple fuels a detailed model of combustion can be complex and would be dependent on temperature, pressure and the precise gas mixture (notably oxygen concentration). Most fire simulations use a simple model where *fuel* and *oxidant* react (depending on temperature and mixing rate) to form a *product*. Other CFD simulations of combustion systems (such as furnaces or internal combustion engines) may use much more complex models that can, for example, include the production of NOX and other minor products. This would be too elaborate for most fire safety simulations where the precise composition of the fuel that is burning at any given time is uncertain (eg piles of luggage in an airport departure lounge). However, where the exact nature of the fuel is known (eg a spill in an industrial process) more complex models may yield useful and reliable additional data.

The results of the simple models can be processed to give estimates of minor products such as carbon monoxide or hydrogen cyanide using empirical relationships of measured yields. These can be used in toxicity calculations but care must be taken to ensure that the empirical relationships relate to the conditions prevailing in the simulation. The complexity of combustion modelling should not be underestimated.

A final simplification is not to include combustion and to assume a fire can be represented as a source of heat. This requires the user to prescribe where the flaming region of the fire is. If too small a volume is used then the temperatures at the fire source will be unrealistically high. This will result in overestimation of the buoyancy and the initial momentum of the fire plume which means that errors may propagate through the flow field. Conversely, if the flame volume is too large then the source temperature may be too low and again the initial flow at the fire source will be incorrect. In addition, flame shape is not easy to predict, flames lean towards walls and become extended under ceilings; these effects will be predicted if a combustion model is used but must be prescribed when the fire is treated simply as a heat source.

4.4.2 Thermal radiation

In fire problems, heat transfer by thermal radiation may be significant as already explained in section 2.2.2. If the main objective of the simulations is to investigate smoke movement, then a flux model may be adequate to account for radiation losses from the hot smoke. The more complex and computationally demanding models such as the *discrete transfer method* will be required for fires approaching flashover and for fire spread problems.

A further complication to the prediction of thermal radiation is that the medium through which it passes may absorb the radiation and, in the case of a flame, emit radiation. CFD models may include this as a simple constant or calculate the constant depending on local concentrations of water, carbon dioxide and soot. This identifies a link between the combustion and radiation models.

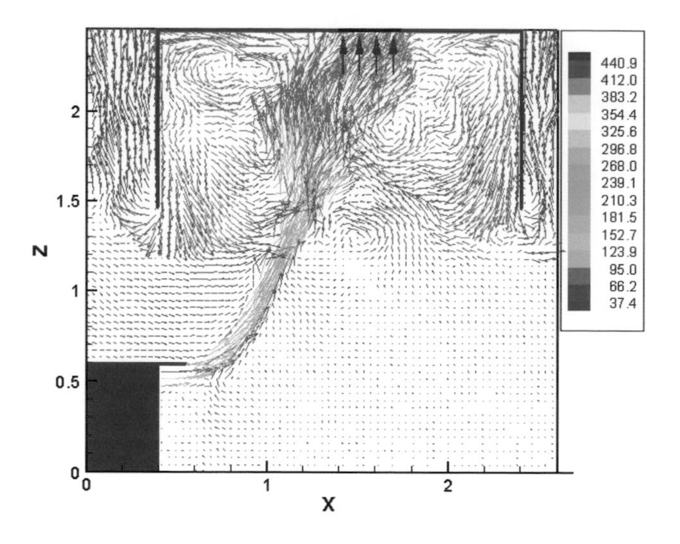

4.4.3 Turbulence

CFD fire simulations typically use a Reynolds-Averaged Navier-Stokes (RANS) or LES turbulence model. These provide data that are used in the combustion models. Simulations with a RANS model will provide time-averaged quantities (and the magnitude of the fluctuations), whereas LES models show the fluctuations and may require post processing to find time-averaged values.

4.4.4 Heat transfer into solid surfaces

The transfer of heat from the smoke and hot gases into the structure of the building may be quite significant. This has two aspects, firstly the cooling of the hot gases, which can reduce their buoyancy leading to contamination at lower levels and secondly heating the building structure which may lead to fire spread, deformation and ultimately structural collapse of the building.

CFD models may include several models for heat transfer into solids.

4.4.4.1 Conduction depth

As the duration of most fires is short relative to the time taken for heat to be conducted through the thickness of the building components, then a simple conduction model that assumes that the material is thermally thick may be used. This assumes that the unexposed face of the object does not change temperature during the fire simulation.

4.4.4.2 Conjugate heat transfer

Not all building elements can be considered to be thermally thick. For example, a steel bulkhead is made of a material with good conductivity and is physically thin. The unexposed face will increase in temperature during a fire and estimating this temperature rise may be the object of the simulation. For these cases, a conjugate heat transfer model may be appropriate. This can be implemented by modifying Equation (1) to remove the convection and source terms leaving the basic conduction equation. However, a solid heat transfer problem will require a much finer grid than for air movement, so construction of the grid may become difficult.

4.4.4.3 Finite element models and structural response

It is also possible to integrate finite element (FE) structural models with CFD models so that the CFD model provides boundary conditions for the structural model. However, this capability is more usually used for mechanical engineering problems (such as engine design) than for fire engineering.

4.5 Fire investigation

In addition to fire safety design, CFD can be used to simulate actual events as part of a fire investigation. The approach is to establish a scenario from the available evidence (eg scene of the incident, photographs/videos, witness statements) and then to compare the results of the simulation with known events. The original hypothesis can then be refined (eg changing the sequence of doors or windows opening/breaking). The process of refining the hypothesis may be used to highlight consistency in witness statements and contribute to building an incident timeline. The simulation results can also be used to estimate conditions at specific locations during the fire and these data may provide information to estimate the performance of different materials or establish conditions required for reconstructions.

It must be noted that a fire simulation will not 'solve' a fire investigation on its own but it may link pieces of evidence and allow the fire investigator to explore the significance of various features of the fire. This may help the fire investigator to justify eliminating some scenarios and to identify the more likely ones.

5 Ventilation and thermal comfort

One of the main uses for CFD in the built environment is to optimise the ventilation system and to predict 'thermal comfort' in a designed building space. As technologies progress, there are two different approaches in the design of a ventilation system for a building:

- to push natural ventilation as far as possible
- to design a complex ventilation system combined with a complex control system that relies on the deployment of unconventional diffusers.

However, for both approaches, designers are seeking to provide a comfortable environment for the building occupants while at the same time reducing energy consumption. By its nature, natural ventilation is less tightly controlled than air conditioning. Furthermore, it is difficult to predict the aerodynamic performance produced from the deployment of unconventional diffusers.

CFD modelling is often used to predict the likely performance of a building design and is able to predict, in considerable detail, complex patterns of air flow and air temperature distribution. Additionally, it enables thermal comfort to be predicted, so designers can directly assess the impact of their design decisions. Consequently, a better understanding and innovative control strategies for indoor environments can be developed, providing effective design tools for improving the thermal performance of the building and occupant comfort and reducing energy consumption.

CFD has been used successfully to predict the ventilation performance of many advanced naturally ventilated buildings. One such example is the Environmental Building at BRE. A case study is presented in chapter 9 showing the design of the building and how CFD modelling was used to demonstrate the effectiveness of the proposed ventilation system.

5.1 Defining indoor thermal comfort

Indoor 'thermal comfort' means that a person wearing a normal amount of clothing feels neither too cold nor too warm. Thermal comfort is important both for a person's well-being and productivity. It can be achieved only when the air temperature, humidity and air movement are within a specified range, often referred to as the 'comfort zone'. There are two empirically based thermal comfort parameters, predicted mean vote (PMV) and the predicted percentage dissatisfied (PPD), as defined in BS EN ISO 7730[14]. These two parameters can be derived from normal CFD results.

The concept of thermal comfort was first proposed by Fanger[15] and it is when the skin temperature and sweat secretion lies within narrow limits. Measurement was carried out in a climatic

chamber, in which sweat rate and skin temperature were measured on people who considered themselves comfortable at various metabolic rates. Thereafter, Fanger expressed the optimal conditions for thermal comfort by the regression line of skin temperature and sweat rate at different metabolic rates in data from these experiments. In this way, an expression for optimal thermal comfort can be deduced from the metabolic rate, clothing insulation and environmental conditions.

Fanger's basic assumption is that thermal comfort is defined in terms of the physical state of the body rather than that of the environment, ie what we actually sense is skin temperature and not air temperature. Thermal comfort requires:

- *Thermal balance*, ie rate of heat loss = rate of heat production
  This is a necessary but not sufficient condition for comfort (eg sweating may lead to heat balance but may not be comfortable).
- *Mean skin temperature*
  This should be at an appropriate level for comfort.
- *Sweating*
  Comfort is a function of a sweating rate, which is also a function of activity and metabolic rate.

PMV is the mean value from a group of people. Fanger extended the PMV to predict the proportion of any population who will be dissatisfied with the environment. A person's dissatisfaction was defined in terms of their comfort vote. Those who vote outside the central three scaling points on the ASHRAE scale[16] were counted as dissatisfied. Hence, PPD is defined in terms of the PMV, and adds no information to that already available in PMV. The distribution of PPD is based on observations from climatic chamber experiments and not from field measurements.

PMV can be predicted from Fanger's equations given in Box 1.

Box 1: Prediction of PMV using Fanger's equations[15]

$$PMV = [0.303 \times \exp(-0.036 \times M) + 0.028] \times \begin{Bmatrix} (M-W) - 3.05 \cdot 10^{-3} \cdot [5733 - 6.99 \cdot (M-W) - p_a] - 0.42 \cdot [(M-W) - 58.15] \\ -1.7 \cdot 10^{-5} \cdot M \cdot (5867 - p_a) - 0.0014 \cdot M \cdot (34 - t_a) \\ -3.96 \cdot 10^{-8} \cdot f_{cl} \cdot \left[(t_{cl}+273)^4 - (\overline{t_r}+273)^4\right] - f_{cl} \cdot h_c \cdot (t_{cl} - t_a) \end{Bmatrix} \quad \text{(Eqn 3)}$$

$$t_{cl} = 35.7 - 0.028 \cdot (M-W) - I_{cl} \cdot \left\{3.96 \cdot 10^{-8} \cdot f_{cl} \cdot \left[(t_{cl}+273)^4 - (\overline{t_r}+273)^4\right] - f_{cl} \cdot h_c \cdot (t_{cl} - t_a)\right\} \quad \text{(Eqn 4)}$$

For $2.38 \cdot |t_{cl} - t_a|^{0.25} > 12.1 \cdot \sqrt{v_{ar}}$ then (Eqn 5)

$$h_c = 2.38 \cdot |t_{cl} - t_a|^{0.25}$$

Otherwise

$$h_c = 12.1 \cdot \sqrt{v_{ar}}$$

For $I_{cl} > 0.078\ \mathrm{m^2K/W}$ then (Eqn 6)

$$f_{cl} = 1.00 + 0.645 \cdot I_{cl}$$

Otherwise

$$f_{cl} = 1.00 + 1.290 \cdot I_{cl}$$

where the terms are:

- *functions of clothing:*
  I_{cl} = clothing insulation
  f_{cl} = ratio of clothed/nude surface area
- *functions of activity:*
  M = metabolic heat production (W/m²)
  W = Metabolic free energy production (external work)(W/m²)
- *environmental variables:*
  t_a = air temperature (°C)
  t_r = mean radiant temperature (°C)
  V_{ar} = relative air speed (m/s)
  P_a = vapour pressure of water vapour (mBar)
  t_{cl} = surface temperature of clothed body (°C)
  h_c = convective heat transfer coefficient

With the PMV value determined, the PPD is calculated using:

$$PPD = 100 - 95 \cdot [\exp(-0.03353 \cdot PMV^4 - 0.2179 \cdot PMV^2)] \quad \text{(Eqn 7)}$$

The environmental variables in Equations (3) to (5) are calculated in a normal CFD simulation. By predefining clothing and activity, and applying Equations (3) to (7), PMV and PPD can be obtained. Furthermore, they can also be presented as contours and animations just like other output parameters from a CFD simulation. Based on the distribution and level of PMV and PDD, the designer can improve and refine a design accordingly.

6 Wind loading

Novel buildings with complex geometries are being constructed. Wind loading codes of practice are limited to simple building geometries and offer little to no guidance on complex geometries. Hence, great care needs to be taken when simple codes are used in complex situations since these can result in unsafe designs. The effects of wind flow over tall and/or geometrically complex structures are often outside the scope of wind loading codes of practice and require detailed wind tunnel investigations to establish overall and local wind pressures. This is particularly true for tall buildings with high slenderness ratios and/or asymmetric plans. Part of the reason for this is that BS EN1991-1-4:2005[17] (and its predecessor BS 6399-2:1997[18]) uses quasi-static methods which are not applicable for buildings susceptible to dynamic loadings.

Scaled-model wind tunnel testing with correct inlet boundary layer simulation is an established industry design tool and is capable of:

- quantifying time-dependent surface pressures, including the complex across wind and torsional loadings
- optimising the orientation of the proposed building.

However, this approach is limited by the need to maintain proportionality between the scaled turbulence characteristics and the scales of the building model, especially if the topography is significant. Additionally, it is essential to ensure Reynolds number effects on the pressures are kept to a minimum. As noted by Sun *et al*[19], a computational approach is more flexible than traditional wind tunnel experiments. It can give a full-scale simulation of a building. Moreover, a fully coupled solution between CFD and finite element modelling (FEM) can also be developed to model the fluid–structure interaction (FSI).

Due to the complex flow interference phenomena induced from buildings, including total turbulent length scales at different regions in the same calculation domain and separation areas larger than the body size of the structure, it has proved difficult for CFD to model external flow around buildings and predict accurate surface pressures for buildings. In addition, all turbulence models were designed to model streamline surfaces. As a result, less work has been performed on predicting time-dependent surface pressures and CFD has not advanced enough to replace wind tunnel testing in this respect. However, it is beneficial when used as a complementary tool to the wind tunnel testing.

6.1 Effect of turbulence models in wind loading calculations

It is common knowledge that there is no single turbulence model which is capable of producing accurate results for all types of problem. Turbulence models have a large impact on both the computational cost and accuracy. Consequently, the choice of an appropriate model is normally decided by:

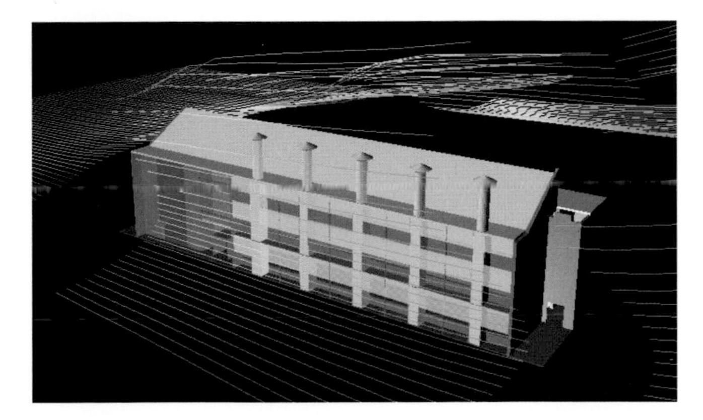

- available computational power
- level of accuracy needed
- amount of time available for the simulations.

6.1.1 RANS

From past experience, mean pressures are generally much better predicted than peak pressures in CFD simulations. This is due to the prevalent use of steady RANS turbulence model. In general, the RANS model modifies the original unsteady Navier-Stokes equations by the introduction of averaged and fluctuating quantities. These equations represent the mean flow quantities only while modelling turbulence effects without a need for the resolution of the turbulent fluctuations. All scales of the turbulence field are being modelled on the assumption that these scales are same length for all directions. As a result, the RANs model gives particularly poor predictions for mounted bluff bodies where separation and reattachment happens often in the flow field.

6.1.2 LES

The LES technique separates large and small scales. The governing equations are obtained by filtering the time-dependent Navier-Stokes equations in the physical space. The filtering process effectively filters out eddies whose scales are smaller than the filter width or grid spacing used in the computations. The resulting equations thus govern the dynamics of the large eddies. Hence, the small scales of turbulence do not require direct modelling and the computational requirement is reduced. However, it requires a significantly finer mesh than traditional RANS simulations and requires a large number of time steps to obtain a stable solution for the flow. Thus, the computational time and required memory are far greater than RANS.

6.1.3 DES

Detached eddy simulation (DES) was developed in an attempt to combine the useful characteristics of both RANS and LES, where RANS is used inside an attached and mildly separated boundary layer, while LES is employed in the massively separated regions. As the model has to identify the different regions, it requires the user to have a detailed understanding of the method and grid quality requirement. Theoretically, DES requires less computational resources than LES.

6.1.4 DNS

The most complete form of CFD is direct numerical simulation (DNS). This technique calculates the direct solution of the Navier-Stokes equations for each control volume. Hence, to capture the complete spectrum of turbulent eddies, the DNS mesh must be smaller than the smallest eddy within the flow. Therefore, the computational cost of DNS is extremely high.

6.1.5 Summary

In conclusion, making use of unsteady RANS, DES and LES simulation techniques is the future of CFD modelling for the prediction of structural loads in wind engineering. However, this requires significant computer resources. Nevertheless, unsteady simulations make it possible to assess the validity of the quasi-steady theory.

6.2 Atmospheric boundary layer

Both wind tunnel testing and CFD methods are sensitive to the boundary conditions which define the simulated atmospheric boundary layer (ABL). The wind speed in the atmospheric boundary layer varies with height and this variation depends on the upstream friction of the ground. Deaves and Harris[20] have derived semi-empirical equations which characterise the known theoretical elements and fitted experimental data accurately. The equation for the equilibrium mean wind profile is:

$$\overline{V} = 2.5u_*\{\ln[(z-d)/z_0] + 5.75(z-d)/z_g - 1.875[(z-d)/z_g]^2 - 4[(z-d)/z_g]^3/3 + [(z-d)/z_g]^4/4\} \qquad \text{(Eqn 8)}$$

where:
u_* = friction velocity and defines the slope of the logarithmic part of the profile.

Three characterised heights are defined in the equation:

Z_g = gradient height, at which the velocity would not be affected by the roughness of the ground and above which the velocity will be constant

d = zero plane displacement height, below which is the interfacial layer

z_o = a height above ground where, theoretically, the wind speed is effectively considered to be zero. This roughness factor is very low in open countryside without trees and vegetation, but rises in city environments.

The 'growing' of an atmospheric boundary layer within a wind tunnel is quite well understood and validated from full-scale measurements. In CFD simulations, this is replaced by Equation (8) which defines the inlet conditions. However, using the equation for the mean wind speed alone is not sufficient to perform a simulation. Gust wind speed (turbulence) plays an important role in wind pressure prediction but is not widely and accurately adopted, particularly for transient turbulence models such as LES. Indeed, LES simulations require the specification of inlet conditions that vary both spatially and temporally. The generation of such conditions is the focus of much current research[21] and represents a real change in the approach to the specification of inlet conditions.

Based on the fact that correlation functions of typical turbulent shear flows have forms not too dissimilar to decaying exponentials, a digital-filter-based generation of turbulence is a suitable technique for LES of spatially developing flows. The artificially generated turbulent inflows satisfy the prescribed integral length scales and Reynolds stress–tensor. At every time step, one set of three-dimensional random data is filtered to generate a set of three-dimensional data with the appropriate spatial correlations. These data are correlated with the data from the previous time step by using an exponential function based on two weight factors. The method developed by Lawson[22] is validated by simulating plane channel flows with smooth walls and flows over arrays of staggered cubes (a generic urban-type flow).

6.3 Domain size

Domain size has an important influence on simulations. To be more specific, for micro-climate and winding load analysis of a single building, the inlet, the lateral boundary and the top boundary should be 5H away from the building, where H is the building height. For buildings with an extension in the lateral direction much larger than the height, the blockage ratio should be lower than 3%. The outflow boundary should be positioned at least 15H behind the building to allow for flow development, as fully developed flow is normally used as a boundary condition.

For the same reason, this outflow length should also be applied for an urban area with many buildings, where H is to be replaced H_{max}, the height of the tallest building. To prevent an artificial acceleration of the flow over the tallest building, the top of the computational domain should be also at least $5H_{max}$, away from this building. For the blockage ratio, the limit of 3% is recommended, although there are no results on whether it is better to include more of the surrounding buildings in the model and reduce the distance of the lateral boundaries from the built area.

The extent of the built area (eg buildings, structures or topography) that is represented in the computational domain depends on the influence of the features on the building or region of interest. In some wind tunnel simulations, for example an area with a radius of 300 metres around the building or place of interest is modelled.

Another consideration for wind tunnel simulations is that a building with height H_n is likely to influence the air field to its distance of about 6–10 H_n. Thus, as a minimum requirement, the simulation domain should cover a region at least 6 times of building height H_n in distance. Computational domain size should adopt the similar rule of thumb wind tunnel simulations uses.

While urban areas usually do not possess any geometrical symmetry, simpler obstacles can be symmetric for certain wind directions. In these cases, the symmetry can be used as a boundary condition and the computational domain can be halved. But it should always be verified in advance that the flow is really symmetric by performing a simulation in the full domain, as even geometric symmetry can produce asymmetric flows.

The general rule for the domain size of CFD simulations is that the boundaries of the simulation domain should not be located in regions where there are recirculating flows.

7 Microclimate around buildings

The microclimate around buildings has become more complex as taller and generally deeper plan buildings have been built over the years. This can cause adverse wind effects at ground level and on balconies, resulting in discomfort and safety concerns for pedestrians. The wind effect on buildings and structures is a complicated problem and is exacerbated when transferred to an urban or city centre environment. CFD is used routinely to assess the wind impact on outdoor thermal comfort. Thermal comfort criteria developed by Tom Lawson[22] are often used.

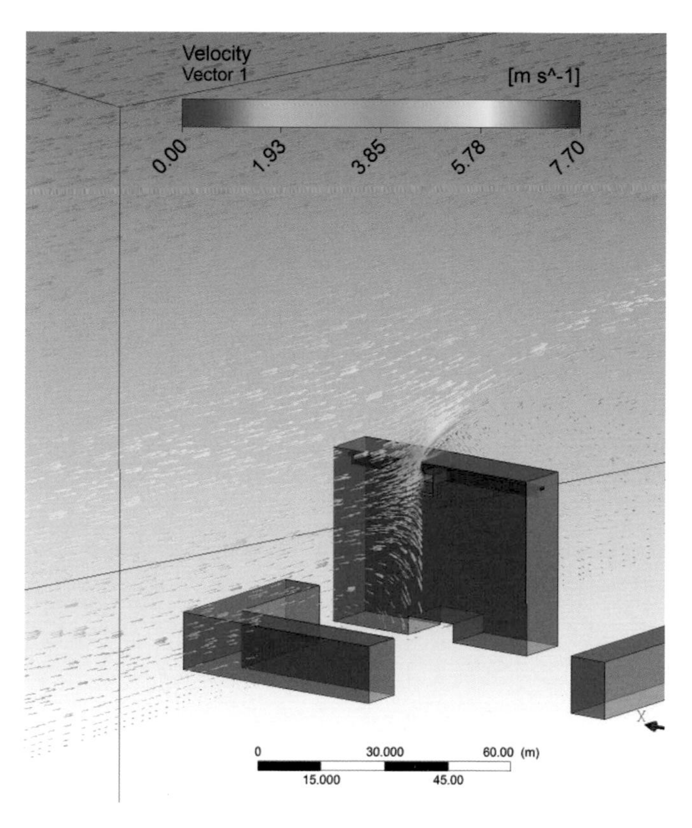

7.1 Identification of potential issues

The elements that will affect the wind environment between buildings include:

- local wind climate
- surrounding areas and local terrain topography
- building form, shape, height, location and orientation
- landscape (trees, fences, hedges)
- effects of the proposed development on the surrounding pedestrian areas
- construction stages of the buildings that form the scheme and surrounding developments.

As there are so many elements, a CFD model can give a holistic analysis of the effects of the characteristics of the local area.

7.2 Wind climate

7.2.1 Advanced boundary layer

The atmospheric boundary layer (ABL) plays an important role for wind climate simulation, as it does for wind loading simulation. The general principles shown in chapter 6 are also applicable for wind climate.

7.2.2 Wind speed variation within urban environments

The description of the vertical wind profile given by Equation (6) is an idealisation, and will rarely be achieved at a specific location in an urban area. Effects such as sheltering by buildings, funnelling, and generation of additional turbulence and vortices, will all combine to generate a more complex pattern of wind speeds than over open country.

When the wind impinges on tall buildings, it tends to 'wash down' the front of the building. The wind speed will increase with building height and so the taller the building, the greater the effective wind speed at pedestrian level. Fairly strong reverse wind speeds can result at pedestrian level.

The existence of vortices between buildings is only significant when the height ratio of the buildings exceeds two. Turbulence created behind buildings has a large effect on pedestrian comfort. The turbulence or gust speeds will increase in the wake of buildings. While the mean wind speed will be lower, the pedestrians may still experience undue discomfort. Turbulence intensity is the ratio of the standard deviation of the fluctuations and the mean wind speed. In an urban environment within the ABL, the turbulence intensity is generally between 0.15 and 0.2. This may increase by 50% behind large buildings. In some areas on the leeward side of the building, the turbulence may increase by up to 30%, despite the hourly mean wind speed being lower.

As the wind is deflected around and between buildings, the mean wind speed is generally accelerated between buildings as they get closer together, which is called funnelling. Lawson[22] stated that the critical or scaling dimension d is either the cross-dimension or twice the height of the building, whichever is smaller. If the gap between the buildings is less than $d/4$ or greater than d, there is no funnelling. The maximum effect occurs when the gap is equal to $d/2$. The highest mean wind speeds are experienced at a distance downstream from the front of the building, equivalent to half the height of the building[23]. This is the mean wind speed experienced at pedestrian height between the buildings. At this distance, the wind speed could be increased by almost 20%. Further downstream, this effect diminishes and in a distance of some 20–30 metres will return to the mean expected wind speeds.

Vortices resulting in increased turbulence are created at all leeward or trailing edges of a building, particularly the roof and the corners of the buildings in the different wind directions. Generally, the velocities in these wake regions will be between 0.8 and 1 of the upstream mean wind speed at building height.

7.3 Criteria for pedestrian comfort

Comfort is generally related to the 'wind chill' factor experienced by the pedestrians and is quite different for vigorous walkers than for sitting pedestrians.

The Lawson criteria form the basis of comfort for pedestrians. They describe the level of dissatisfaction experienced by pedestrians. The Beaufort scale[24] is the standard metric for categorising wind speed and is used to measure the percentage of hourly mean wind speed throughout the whole year. This scale gives a range of 12 wind speed categories into which each band has a range of hourly mean wind speeds. Larger numbers indicate higher speed with B12 being the highest wind speed in the scale. The comfort criteria levels are described in Table 1. Different activities have different requirements based on the accumulated percentage of time for which the wind has a higher speed than a certain Beaufort scale. From Table 1 it can be seen that there are six scenarios. Taking the 'sitting' scenario as an example, if the Beaufort scale B3 happens more than 1% of the time, the condition in that location is considered to be unacceptable.

To determine the wind flow patterns around the buildings and predict wind speeds at vulnerable locations, CFD simulations with different wind directions are required.

Table 1: Lawson comfort criteria

Scenario description	Level	Threshold of wind speed	
		Unacceptable	**Tolerable**
Roads and car parks	A	6% > B5	2% > B5
Business walking	B	2% > B5	2% > B4
Pedestrian walk-through	C	4% > B4	6% > B3
Pedestrian standing	D	6% > B3	6% > B2
Entrance doors	E	6% > B3	4% > B2
Sitting	F	1% > B3	2% > B2

7.4 Outdoor thermal comfort criteria

The quality of outdoor thermal comfort affects people's behaviour and use of outdoor spaces. Responses to the microclimate may be unconscious, but they often result in a different use of open spaces in different climatic conditions. Successful outdoor spaces attract large numbers of people, generate cultural diversity, enhance economic vibrancy and provide places for people to meet and socialise.

Research into outdoor thermal comfort is relatively new and has to address a number of highly variable parameters that include user groups, activities and climate. Similar to the indoor model, a heat balance approach to outdoor thermal comfort is used to develop outdoor indices, with modifications made to account for solar radiation. Physiological equivalent temperature (PET)[25] and outdoor standard effective temperature (OUT_SET)[26] are two examples of indices that are widely used to quantify outdoor comfort.

Case studies

8 Case study 1 Fire simulation

In 1982 the National Institute of Standards and Technology (NIST), conducted a series of 55 full-scale fire experiments in a small compartment using different combinations of fire size, position and opening configuration[27]. The fire was run for 30 minutes before measurements were made so that the compartment reached a steady state. These tests were extensively instrumented and have become a valuable data resource for the validation of various computer fire simulation programs. One of the cases is used here to illustrate various aspects of conducting a fire simulation using CFD.

8.1 The test scenario

The test compartment was 2.18 m high and had a square plan with sides 2.8 m long. The test being considered here (Test 12) had a door 0.5 m wide and 1.83 m high on one wall and a fire in the centre with a heat release rate of 62.9 kW. This test was selected because it was repeated and in both cases the measured inflow and outflow were very close indicating a precise set of measurements. The building was constructed from lightweight concrete covered with ceramic blanket on the inside. The fire was a methane burner 0.3 m in diameter. The tests were instrumented with a number of thermocouple columns inside the compartment and an extensive measurement grid (temperature and velocity) so that an estimate of the mass flow into and out of the compartment could be made.

8.2 Setting up the problem

Figure 12 shows the geometry created with the BRE JASMINE software[28]. The fire source is square with an area equivalent to the NIST 0.3 m burner (0.26 m on each side). The thermal properties of the walls are taken to be those of the ceramic blanket. The values are given in Table 2.

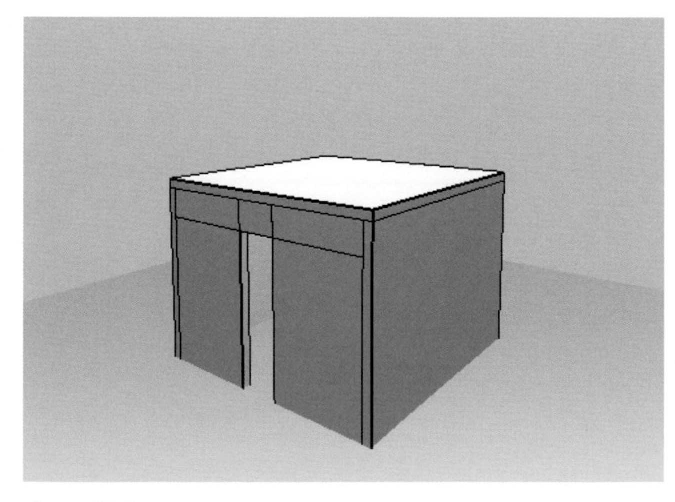

Figure 12: Compartment geometry

Table 2: Thermal properties of walls

Wall property	Value
Density	2700 kg/m^3
Thermal conductivity	0.1 W/m/°K
Specific heat capacity	1130 J/kg/°K

The material properties in Table 2 are 'typical' values for a generic product. In practice, the values will be dependent on temperature.

8.3 Simulations

8.3.1 Run 1

A calculation domain around the geometry was defined. This extended 1.5 m outside the door and 2 m above the roof to ensure the flow at the doorway was not influenced by the fixed pressure boundary condition specified outside the door, the sides and top of the domain. The floor of the domain was a solid material with no heat losses. A simple grid with cell dimensions between 0.1 m and 0.2 m was used for run 1. No attempt was made to refine the grid around the fire or the door way although an increased number of vertical calls are provided over the expected flame height and there is a cell located close to each wall. Figure 13 shows the grid and extent of the calculation domain.

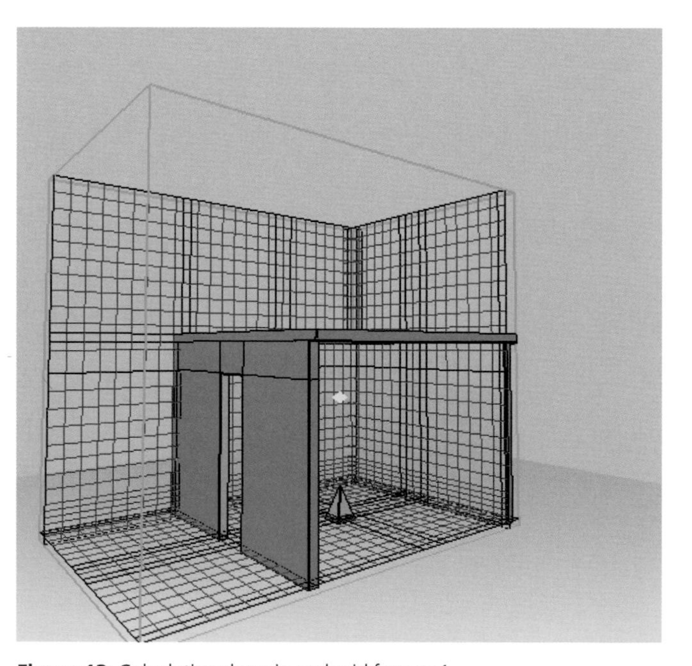

Figure 13: Calculation domain and grid for run 1

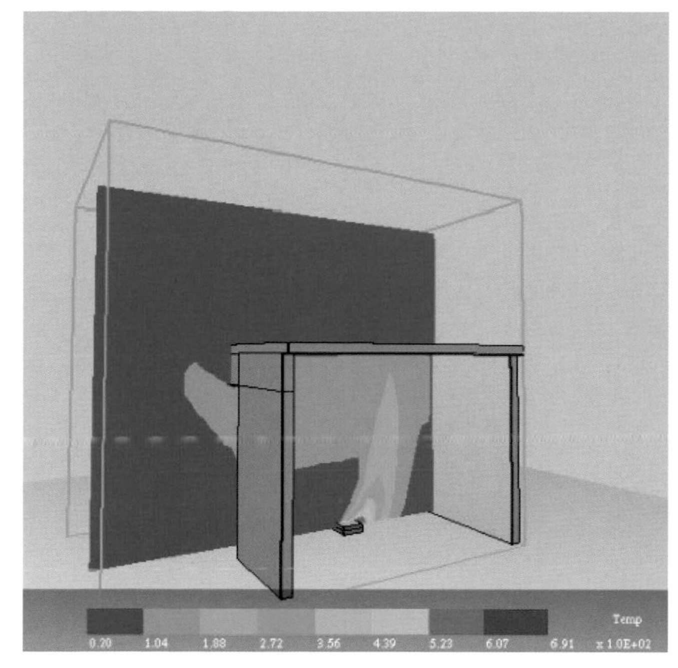

Figure 14: Temperature contours for run 1

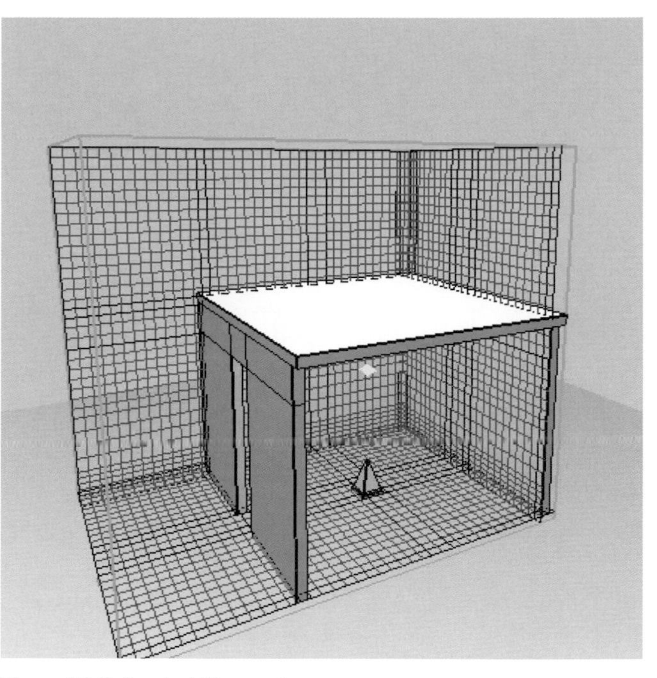

Figure 15: Refined grid for run 2

The simulation used a total of 16 240 cells.

The simulation was run with JASMINE using the six flux (P1) radiation model. Using a 3 GHz dual core PC a converged solution (mass and energy balances greater than 99%) was reached in approximately 8 minutes. A contour plot showing temperature on the centre line is shown in Figure 14.

Figure 14 shows a warm layer under the ceiling of the compartment with hot gases spilling out through the doorway. However the maximum temperature calculated is 690°C which is outside the range of temperatures that would be expected in a fire (800–1200°C) and the mass flow rate through the door is 0.4 kg/s compared with 0.46 kg/s measured in the experiment. The solution seems to be qualitatively reasonable. However, it is not quantitatively accurate.

8.3.2 Run 2

To improve the accuracy it was noted that the fire plume leans over towards the back of the compartment. The grid was refined so that the number of cells was increased to 38 504 with an average cell size of about 0.12 m. Figure 15 shows the refined grid for run 2.

Run 2 took 22 minutes to reach a converged solution. Figure 16 shows the temperature contours (using the same scale as Figure 14) where the maximum temperature is still approximately 690°C. However, the flow through the doorway was 0.43 kg/s compared with the measured value of 0.46 kg/s, which is within the experimental accuracy of 10%.

Closer inspection of the grid shows that the number of cells near the fire did not change significantly between runs 1 and 2 therefore some local grid refinement was conducted for a third run. As well as adding cells near the fire source,` some cells were removed near the boundaries outside the room where there was little air movement.

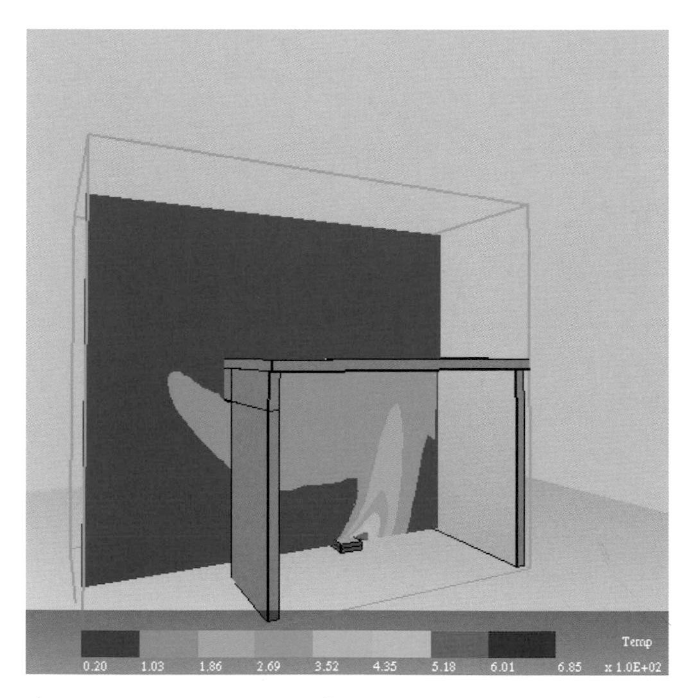

Figure 16: Temperature contours for run 2

8.3.3 Run 3

The grid for run 3 used 42 656 cells. The predictions showed a maximum temperature of 970°C and a flow rate of 0.42 kg/s through the doorway. Figure 17 shows the predictions of gas velocity and temperature in the doorway from JASMINE compared with the measured values (Test 12 and repeat Test 612). The predictions are within the spread of the experimental measurements but the difference in the velocity between the prediction and measurement at the top of the door indicates that there may be some mixing process occurring (such as eddy formation) that is not being captured by the simulation.

8.3.4 Run 4

The grid was refined further at the top of the door to give a total 54 144 cells and bring the computation time to 35 minutes. The predicted and measured velocity profiles at the door are shown in Figure 18.

The predictions and measurements are now very close with the exception of the temperature of the gases entering the compartment and the velocity at the very top of the doorway. The temperature difference is due to variations in ambient temperature. The velocity difference may be due to details on the top edge of the door which were not included in the simulation. This may also be why the simulations have a slightly lower (although within the experimental accuracy) prediction of mass flow through the doorway.

8.4 Summary

This case study shows that a global refinement of the grid may not be sufficient to obtain a good prediction. Examination of the temperature predictions showed that the temperature in the fire was not realistic and a local refinement of the grid was required to obtain a reasonable value. Local grid refinement at the top of the door also improved the doorway temperature and velocity profiles.

When comparing the results of CFD simulations with experimental measurements, the accuracy of the measurements should also be considered.

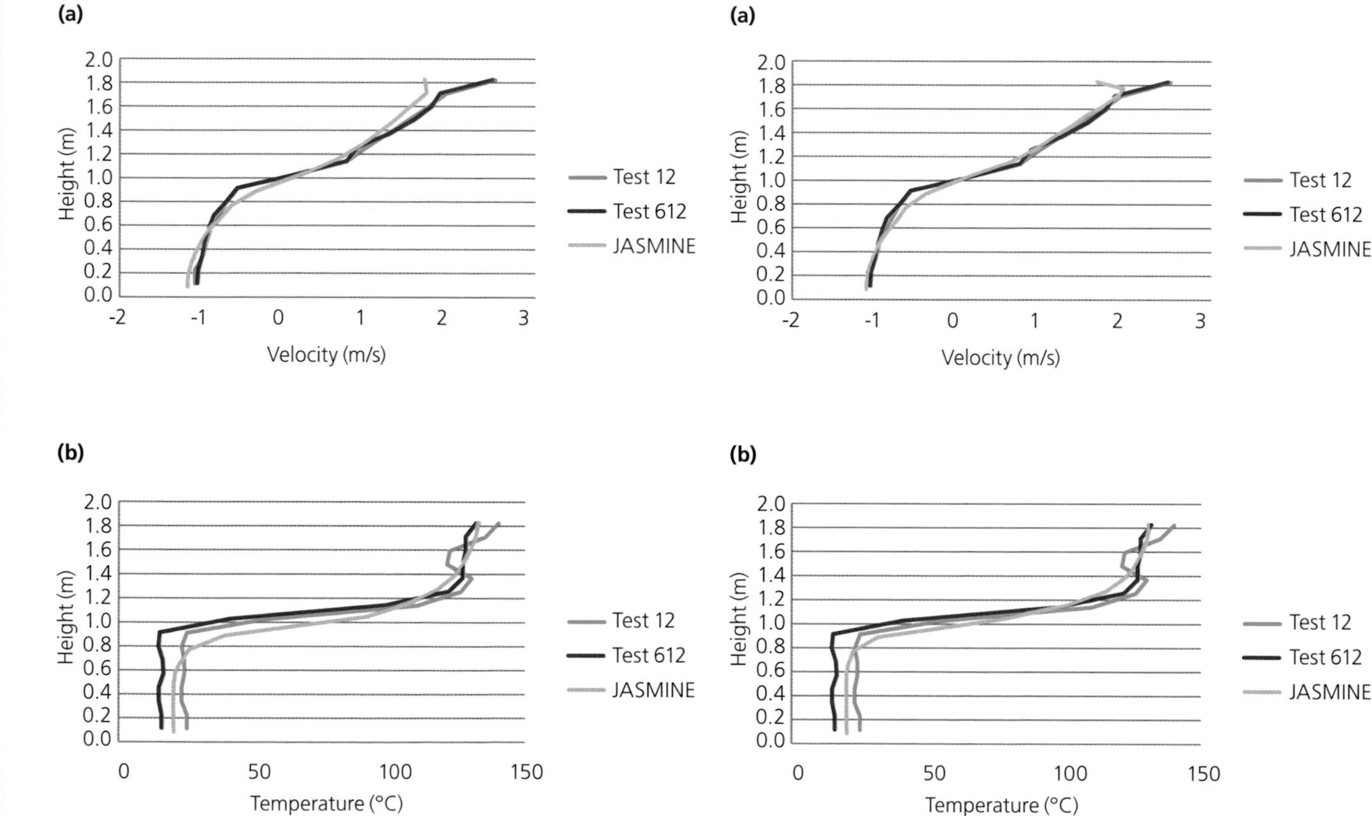

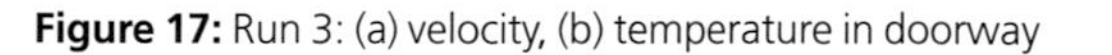

Figure 17: Run 3: (a) velocity, (b) temperature in doorway

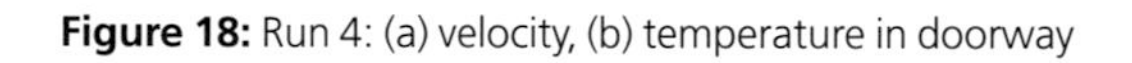

Figure 18: Run 4: (a) velocity, (b) temperature in doorway

9 Case study 2 Natural ventilation

CFD simulations have been conducted to assess the performance of the novel natural ventilation system used in the Environmental Building at BRE's Watford site on a warm summer day with a light breeze.

9.1 The building

The BRE Environmental Building does not have air conditioning, but the building design includes a novel combination of cross-ventilation, a precast concrete wave ceiling with cavity air channels and five external ventilation shafts to provide natural ventilation. The ground- and first-floor office spaces are 30 m long (east–west) by 13.5 m wide (north–south), with a 3.3 m distance from the floor to the ceiling crest.

Nine cavity ceiling channels run in the north–south direction, each channel consisting of separate north and south sections opening into the office space approximately 5 m from the north façade. Whereas all north sections open, via building management system (BMS) controlled windows, to the external atmosphere, alternate south sections open into external ventilation shafts (again under the control of the BMS). Figure 19 shows the ventilation paths in a cross-section containing an external shaft, and shows also the division of the office space between an open-plan area and cellular offices. Figure 20 shows, from inside the office, the air paths to a ventilation shaft, which, in addition to the ceiling channel, includes two low-level 'hopper' windows which open directly from the open-plan area.

Additional office ventilation may be provided by high-level windows that open to the outside at the ceiling crests midway between the channels, and from a provision of manually controlled low-level windows. With the doors closed, the cellular offices are effectively isolated from the main office space, with 'single-sided' ventilation provided locally by windows in these offices. As the subject of a European Commission THERMIE project, further information about BRE's Environmental Building and its conceptual design can be found in references [29] and [30].

9.2 The CFD model

The appropriateness of the CFD model for a building depends not only on the characteristics of the fluid flow and related processes, but also the geometrical aspects of the problem. Although many buildings may be formed from simple rectangular shapes, others include more complex shapes (such as the office ceiling inside the Environmental Building) which require accurate numerical representation. The accuracy of the CFD solution relies heavily on the quality of the computational mesh on which the discretised equations are solved. Therefore, in buildings with interconnected zones and geometrical complexities it is desirable to employ a CFD model which is capable of representing a complex, three-dimensional geometry using a body fitted numerical mesh to fit irregular boundaries. An unstructured mesh CFD model was used for the study, allowing the geometry to be discretised with an irregular arrangement of tetrahedral and prism elements in an efficient manner.

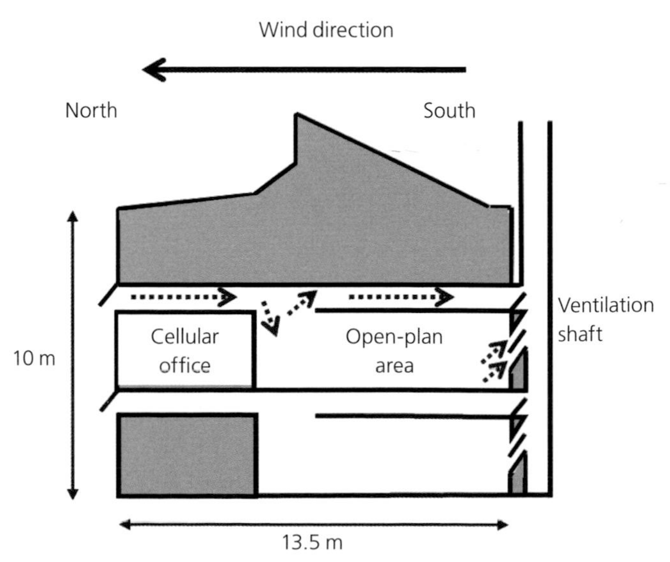

Figure 19: Building cross-section showing ceiling channels and ventilation shaft

Figure 20: Airflow paths to a ventilation shaft

9.3 The problem considered

On a warm summer day with a light breeze, the external wind may be too weak to provide sufficient cross-ventilation and this is the primary reason for the external ventilation shafts being designed to enhance air movement within the office space. The ventilation shafts also have a role to play during the night-cooling operation, again when the external wind is weak.

Three separate mechanisms, two passive and one active, can assist the flow of air from the office into a ventilation shaft, and then into the external environment:

- pressure-induced flow generated by the movement of external air across the top of the shaft
- buoyancy-induced flow generated by solar heating of the glass panels on the front (south face) of the shaft
- a low-speed fan installed near the top of the shaft.

The low-speed fan operates only when the passive measures provide insufficient upward flow. This could occur, for instance, when there is no solar gain and no wind. However, the case studied here is where there is a small airflow across the top of the shafts and the fans are not operating.

The interaction between the internal and external conditions makes it preferable for the flow domain to include the external environment. Ideally, the model should include the whole building as well as the influence of neighbouring structures. This, however, would require either a prohibitively large numerical mesh or a rather coarse one.

To investigate the interaction of the ceiling channels and the ventilation shafts, a reduced three-dimensional slice was modelled. While not capturing all the features of the flow, the office space is sufficiently periodic in the east–west direction for the simulations to capture the main characteristics of the selected scenarios and allow the main parameters, and their interaction, to be investigated.

The simulated domain was extended well beyond the building envelope, where inlet (wind) and free boundary conditions were imposed. The windows opening from the ceiling channels to the outside were included in the model, and were opened to 45°. A 470,000-element mesh was used, with a finer resolution inside the office and ventilation shaft and a coarser resolution elsewhere. Figure 21 illustrates part of the surface mesh inside the office, showing the ceiling slab, the window and part of the office furniture.

In a naturally ventilated building, the time-dependent interaction between the external conditions, the building fabric and the internal thermal load means the boundary conditions vary with time. In practice, however, CFD boundary conditions are often approximated as steady-state, based on the analytical, experimental or assumed values associated with a given time

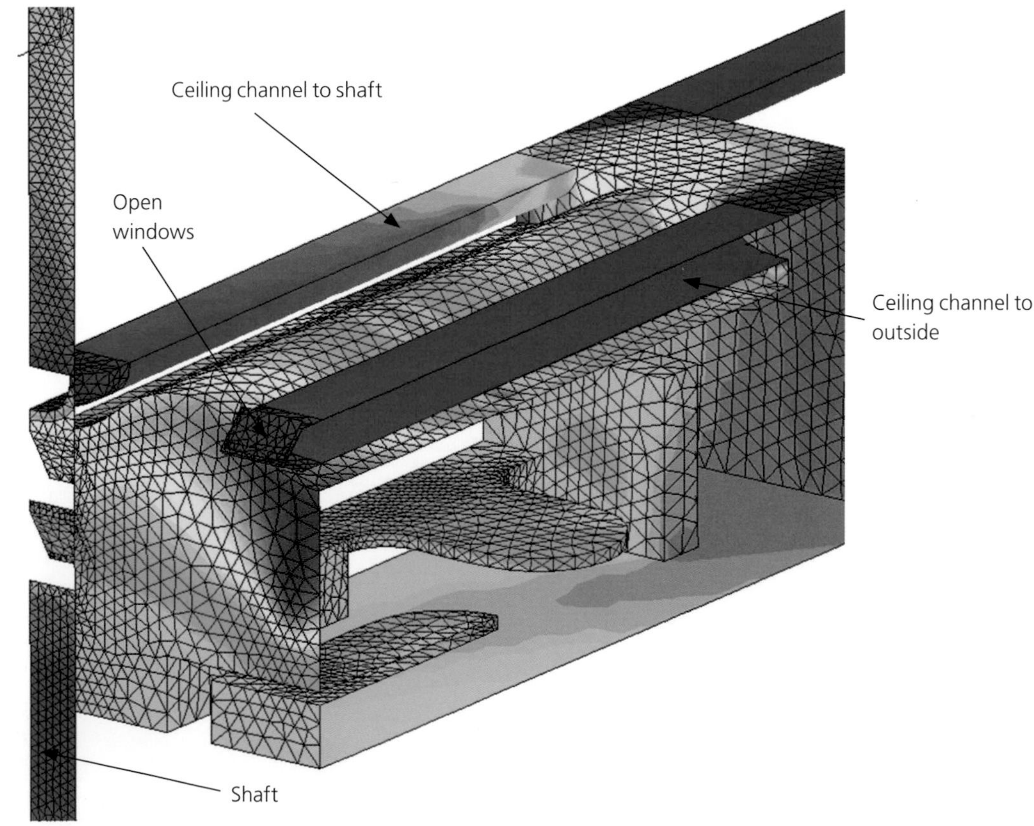

Figure 21: Part of the surface mesh inside the office

of day. In this study, several steady-state conditions were imposed by simulating different external wind speeds, ambient temperatures, heat loads and control strategies.

The standard k-ε turbulence model was employed and all solid surfaces were treated as hydro-dynamically smooth. Although radiation was not modelled directly, its effect can be considered to be included implicitly in the thermal boundary condition at the ceiling slab. A full matrix of simulations was performed, covering the various combinations of boundary conditions. This allowed the sensitivity of the boundary condition parameters to be assessed.

9.4 Summary

The analysis found that the BRE's Environmental Building can perform effectively. Provided there is some breeze, the ventilation stacks were shown to operate effectively even in the absence of any solar heating. This supports the design assumption that in most situations the pressure drop at the top of the stacks should draw air through.

The overall ventilation rate inside the office was correlated to the magnitude of the breeze. At the higher wind speed of 1.5 m/s, in particular, the predicted ventilation rate (air changes per hour, ACH) was quite high at about 7 ACH^{-1}. At the lower wind speed of 0.5 m/s the value was reduced to about 4 ACH^{-1}. These values could be reduced by partially closing the ceiling channel and hopper windows.

On hot summer days, with an outside temperature in excess of 25 °C, the findings indicate that the building would need to operate in a trickle ventilation mode and activate the groundwater cooling system (as happens in practice).

10 References

1. Cox G. Fire modelling. BRE DG 367. Bracknell, IHS BRE Press, 2004.
2. Kumar S. Fire modelling with computational fluid dynamics. BRE DG 511. Bracknell, IHS BRE Press, 2009.
3. Blackmore P. Wind microclimate around buildings. BRE DG 520. Bracknell, IHS BRE Press, 2011.
4. OpenCFD. OpenFOAM CFD software. Visit www.openfoam.com.
5. McGrattan K, Hostikka S, Floyd J et al. Fire dynamics simulator (Version 5), Technical reference guide. NIST Special Publication 1018-5. Washington DC, US Department of Commerce, 2009.
6. Versteeg H K and Malalasekera W. An introduction to computational fluid dynamics: the finite volume method. 2nd edition. Harlow, Pearson/Prentice Hall, 2007.
7. CHAM. PHOENICS CFD software. Visit www.cham.co.uk.
8. Software.Informer. ANSYS CFX software. Visit www.ansys-cfx.software.informer.com.
9. BSI. Fire safety engineering. Assessment and verification of mathematical fire models. BS ISO/TR 13387-3:1999. London, BSI, 1999.
10. United States Nuclear Regulatory Commission (US NRC). Verification and validation of selected fire models for nuclear power plant applications. Volume 1: Main Report. NUREG 1824. Washington DC, US NRC, 2007.
11. United States Nuclear Regulatory Commission (US NRC). Verification and validation of selected fire models for nuclear power plant applications. Volume 7: Fire dynamics simulator (FDS). NUREG 1824. Washington DC, US NRC, 2007.
12. BSI. Application of fire safety engineering principles to the design of buildings. Code of practice. BS 7974:2001. BSI, London, 2001.
13. BSI. Components for smoke and heat control systems. Code of practice on functional recommendations and calculation methods for smoke and heat control systems for covered car parks. BS 7346-7:2013. BSI, London, 2013.
14. BSI. Ergonomics of the thermal environment. Analytical determination and interpretation of thermal comfort using calculation of the PMV and PPD indices and local thermal comfort criteria. BS EN ISO 7730:2005. BSI, London, 2005.
15. Fanger, P O. Thermal comfort: analysis and applications in environmental engineering. New York, McGraw-Hill, 1972.
16. ASHRAE Technical Committees. Thermal comfort. In: ASHRAE Handbook – Fundamentals. Atlanta GA, ASHRAE Inc., 2005.
17. BSI. Eurocode 1. Actions on structures. General actions. Wind actions. BS EN 1991-1-4:2005+A1:2010. BSI, London, 2010.
18. BSI. Loading for buildings. Code of practice for wind loads. BS 6399-2:1997. BSI, London, 1997.
19. Sun D, Owen J S and Wright NG. Application of the k-ω turbulence model for a wind-induced vibration study of 2D bluff bodies. Journal of Wind Engineering and Industrial Aerodynamics, 2009, 97, 77–87.
20. Deaves D M and Harris R I. A mathematical model of the structure of strong winds. Issue 76 of CIRIA report. London, CIRIA, 1978.
21. Zheng-Tong Xie and Castro I P. Efficient generation of inflow conditions for large eddy simulation of street-scale flows. Flow Turbulence and Combustion, 2008, 81, 449–470.
22. Lawson T V. Building aerodynamics. London, Imperial College Press, 2001.
23. Cook N J. The designer's guide to wind loading of building structures. Part 1: Background, damage survey, wind data and structural classification. Watford, BRE, 1985.
24. Met Office. Beaufort scale. Visit www.metoffice.gov.uk/weather/marine/guide/beaufortscale.html.
25. Hoppe P. The physiological equivalent temperature – a universal index for the biometeorological assessment. International Journal of Biometeorology, 1999, 43, 71–75.
26. Gagge A P, Stolwijk J A J and Nishi Y. An effective temperature scale based on a simple model of human physiological regulatory response. ASHRAE Transactions, 1971, 77, 247–257.
27. Steckler K D, Quintiere J G and Rinkinen W J. Flow induced by a fire in a compartment. NBS IR 82-2520. Washington DC, US Department of Commerce, 1982.
28. Cox G. The mathematical modelling of fires in enclosures. Interflam 1985. Proceedings of 6th International Conference on Fire Science and Engineering, 26–28 March 1995, Cambridge. London, Interscience, 1985, pp 56–63.
29. Cook J. Toward a sustainable office building. Solar Today, 1998, July/Aug.
30. Spring M. BRE low-energy offices. Building, 1999, 24, 46–51.